I0605688

There is a flower called evening primrose. It's tall, a bit spindly, and has bright yellow petals. You might have heard of this species because the oil extracted from it is used to treat various human ailments. Many plants are used in this way, and it's likely that there are still more remedies to be discovered from other species of plants. So that is good, but not amazing. What is amazing about evening primrose is that it can hear . . .

It obviously doesn't have ears! But its bell-shaped flowers have evolved to collect and focus a very specific sound into the heart of the flower. And that sound is the buzzing of bees. You must remember that sound is essentially vibrations passing through the air, so perhaps the flower is feeling, instead of hearing in the way that we hear. But why has it become sensitive to the bees' buzz? Well, when it "hears" the bee, it produces more nectar, so the bee gets busy drinking the sweet fluid and . . . gets the plant's pollen all over it, pollen that the bee then transfers to another evening primrose flower, and . . . BINGO—the plant is pollinated! And that is very definitely amazing, and it could be happening in your garden.

You see, we make a big mistake if we think that only animals are interesting, clever, cunning, or important. Plants are so fascinating, so remarkable and, as this book so brilliantly explains, very, very important. So we must learn to admire and love our plant life and also to protect it. Many species are in danger of extinction—forests are felled, bogs are drained, meadows are plowed—all at a terrible risk to our world's health. Nurturing plants will be central to your future, so read on and prepare to be amazed by our wonderful Green Planet.

—CHRIS PACKHAM

OUR GREEN PLANET

BEACH LANE BOOKS
An OUR PLANET book.
An imprint of Simon & Schuster Children's Publishing Division
1230 Avenue of the Americas, New York, New York 10020
For more than 100 years, Simon & Schuster has championed authors and the stories they create. By respecting the copyright of an author's intellectual property, you enable Simon & Schuster and the author to continue publishing exceptional books for years to come. We thank you for supporting the author's copyright by purchasing an authorized edition of this book.
No amount of this book may be reproduced or stored in any format, nor may it be uploaded to any website, database, language-learning model, or other repository, retrieval, or artificial intelligence system without express permission. All rights reserved. Inquiries may be directed to Simon & Schuster, 1230 Avenue of the Americas, New York, NY 10020 or permissions@simonandschuster.com.
First published in Great Britain in 2022 by Puffin Books as *The Green Planet*
Published under license from Children's Character Books Ltd, a joint venture company between Penguin Books Limited and BBC Worldwide Limited
Text © 2022 by Children's Character Books Ltd
Illustration © 2022 by Kim Smith
Introduction © 2022 by Chris Packham
All rights reserved, including the right of reproduction in whole or in part in any form.
BEACH LANE BOOKS and colophon are trademarks of Simon & Schuster, LLC.
For information about special discounts for bulk purchases, please contact Simon & Schuster Special Sales at 1-866-506-1949 or business@simonandschuster.com.
Simon & Schuster strongly believes in freedom of expression and stands against censorship in all its forms. For more information, visit BooksBelong.com.
The Simon & Schuster Speakers Bureau can bring authors to your live event. For more information or to book an event, contact the Simon & Schuster Speakers Bureau at 1-866-248-3049 or visit our website at www.simonspeakers.com.
The text for this book was set in Century Gothic Pro.
The illustrations for this book were rendered digitally.
Manufactured in China
1025 SCP
First Beach Lane Books Edition February 2026
2 4 6 8 10 9 7 5 3 1
Library of Congress Cataloging-in-Publication Data
Names: Stewart-Sharpe, Leisa author | Smith, Kim, 1986– illustrator
Title: Our green planet / Leisa Stewart-Sharpe and Kim Smith.
Description: New York : Beach Lane Books, 2026. | Series: Our planet | Audience: Ages 4–8 | Audience: Grades 2–3 | Summary: "An introduction to the hidden realm of plants that fuel our world"—Provided by publisher.
Identifiers: LCCN 2025007193 (print) | LCCN 2025007194 (ebook) | ISBN 9781665972901 (hardcover) | ISBN 9781665972918 (ebook)
Subjects: LCSH: Plants—Juvenile literature | Plant ecology—Juvenile literature | Human-plant relationships—Juvenile literature
Classification: LCC QK49.S749 2026 (print) | LCC QK49 (ebook) | DDC 580—dc23/eng/20250625
LC record available at https://lccn.loc.gov/2025007193
LC ebook record available at https://lccn.loc.gov/2025007194

BBC

OUR GREEN PLANET

LEISA STEWART-SHARPE
Illustrated by KIM SMITH

Beach Lane Books
New York Amsterdam/Antwerp London
Toronto Sydney/Melbourne New Delhi

THE GREEN PLANET

Around five hundred million years ago, long before dinosaurs roamed, primitive plants crept across this barren rock called Earth. Tiny mosses and liverworts hugged the ground, creating the first soil and pumping oxygen into the atmosphere. This planet became a Green Planet.

Today Earth is still dominated by plants, which outweigh all other life—from tiny duckweed floating in our ponds to giant sequoia trees. Although it's easy to take plants for granted, we depend on these light-eaters, oxygen-generators, and rainmakers for every breath of air and mouthful of food.

With our lives so entwined, we need to understand how our Green Planet grows. In the world of plants, it can feel like time passes more slowly—it can take weeks to unfurl a single leaf. But with patient filming, we can speed up months into minutes and peer into their hidden world.

Be warned: Far from peaceful, it's often a battleground.

Plants may not have brains, but they are "intelligent," and as sophisticated as animals, even going so far as to "trick" animals into working for them. They care for other plants and have the sensitivity to smell, taste, touch, hear, and unbelievably . . . "TALK."

So let's explore our Green Planet: a secret world of plants that's more wonderful than you could ever have imagined.

LIFE-GIVING PLANTS

LIGHT-EATING

OAK TREE

Every day, all around us, plants are working their magic through **PHOTOSYNTHESIS**, which turns sunlight into food.

SUNLIGHT is absorbed through the plant's solar-panel leaves.

CARBON DIOXIDE is taken from the air through tiny holes in the leaves (called **STOMATA**).

WATER is pumped through the plant from the roots to the leaves.

OXYGEN is released.

The leaves get to work turning these ingredients into sugar—the food that helps the plant to grow. Some of the sugar is used right away, and the rest is stored in the plant's leaves, roots, or fruits for later.

It's through **PHOTOSYNTHESIS** that plants can release **OXYGEN**.

TRUNK

. . . up the trunk . . .

Water is absorbed through the roots . . .

ROOTS

OXYGEN-GENERATING

During photosynthesis, oxygen is released back into the air through the leaves' stomata. Incredibly, it takes about eight trees to provide one human with the oxygen they need for a year.

RAIN-MAKING

An adult tree can take in 50,000 liters of water a year, which is around twenty-seven times more water than a fire engine can carry! But it's not all stored—the plant uses the water to send nutrients up through the plant and to cool down its leaves. When the plant is cooling down, water evaporates through the leaf stomata in a process called **TRANSPIRATION**.

This evaporated water floats up into the sky and forms clouds. When the clouds meet cool air, the water condenses into tiny droplets. As these droplets get bigger, they fall back down to Earth as rain, hail, sleet, or snow, completing the water cycle.

Plants in the Amazon rainforest release so much water they create an invisible river in the sky—larger than any river on land.

All around us, plants are busy taking in light, breathing out oxygen, and releasing water.

But that's not all they're up to . . .

TROPICAL WORLDS

THE BATTLE FOR LIGHT

This bustling tropical rainforest is quietly ruled by plants. A few giant trees tower above the unbroken canopy, peaceful among the clouds. But far below, on the dark forest floor . . . it's a *battleground.*

Even when the midday sun is high overhead, only a few shards of light pierce the canopy like tiny spotlights. In some places, less than 5 percent of sunlight reaches the forest floor.

Without light, growth is slow, so plants have learned to fight for their moment in the sun. Climbers and creepers lash around trees as they climb up, up, up, to where air plants called **EPIPHYTES** have created a world of their own—dangling, roots and all, above the ground.

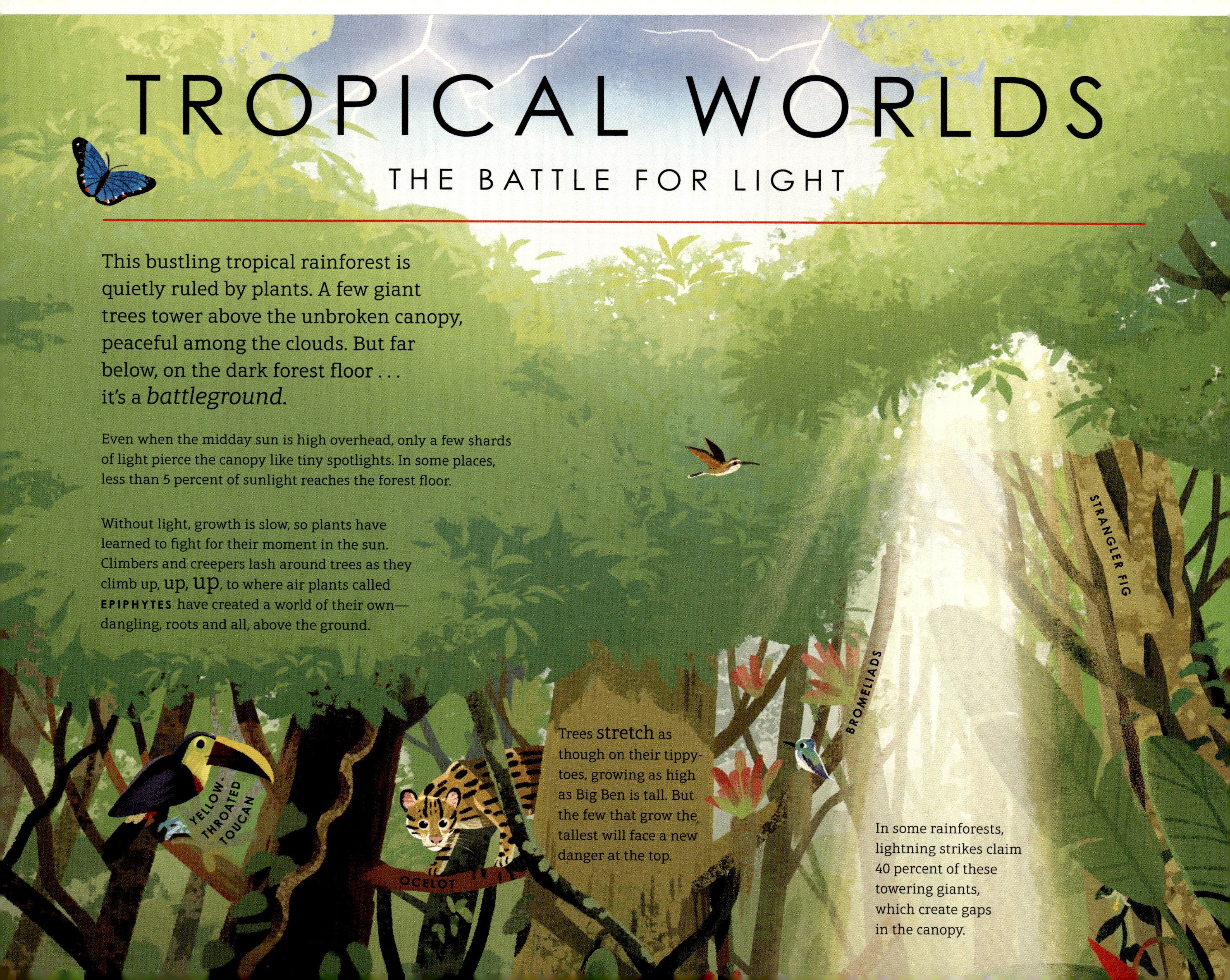

Trees stretch as though on their tippy-toes, growing as high as Big Ben is tall. But the few that grow the tallest will face a new danger at the top.

In some rainforests, lightning strikes claim 40 percent of these towering giants, which create gaps in the canopy.

STRANGLER FIG

BROMELIADS

YELLOW-THROATED TOUCAN

OCELOT

Light

breaks

through . . .

and it's what the forest floor plants have been waiting for.

Suddenly, the race is on, as even the slowest growers **unfurl** and **climb** toward the sun. Young trees that have taken a decade to reach a stunted size now bolt for the sky, as seeds burst from the ground all around them.

Competition is fierce in their battle to reach the light.

STORIES
FROM THE TROPICAL WORLDS

THE SPEEDY SNACK STOP

LOOK . . . a gap in the forest. Here among Costa Rica's stately giants, an old tree has toppled, providing a chance for this **BALSA TREE**. Spindly and admittedly a bit scruffy, it's in a race against time.

A seedling shoots up at dizzying speeds, and by the time it's ten, it could stretch up to eighty-eight feet high (that's twice as tall as a brachiosaurus). It's definitely not the biggest, or even the sturdiest, tree in the forest.

But that's all part of the plan.

For the balsa, timing is everything. When all the other plants have slowed down for the dry season, the balsa opens its afternoon blooms—its branches sagging under the weight of its flashy flowers. These flowers are filled with rich, syrupy nectar, and they stay open all night. Soon it's the busiest snack stop around.

Here come a couple of thirsty regulars: the shy **KINKAJOU** and **DERBY'S WOOLLY OPOSSUM**. They've both arrived for a hefty shot of nectar.

The animals move from flower to flower, draining the huge blooms—their furry faces taking the balsa's pollen with them, pollinating as they go.

It's a feverish pace of life as the balsa grows fast and falls fast,

for a time the forest's favorite snack stop.

THE DROOL POOL

Most crafty **PITCHER PLANTS** know exactly how to rustle up dinner.

These jug-shaped death traps smell irresistible to unsuspecting bugs, which soon slide off the slippery rim into the deadly drool pool below.

With the trap set, all the pitcher needs to do is wait . . .

1 **ANT** scurries by

2 Ant slips

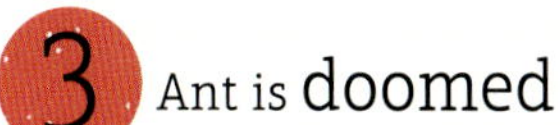

3 Ant is doomed

But on the rain-drenched slopes of Borneo's highest peak, there are fewer bugs to lure in for lunch. And up here, the soil is low in **NITROGEN**, which is needed to help plants grow.

This monster plant, the **KINABALU PITCHER PLANT**, has found an unusual solution to the problem. It's going to get its nitrogen elsewhere.

And here comes its dinner now.

The sweet-toothed **MOUNTAIN TREE SHREW** cautiously approaches the pitcher's gaping mouth and perches on the rim to lick the plant's sugary lid.

Will it fall into the drool pool?

Its bottom dangles over the edge . . .

PLOP!

The shrew poops.

And that's just what the Kinabalu's been waiting for. Shrew poop may not be to everyone's taste, but it's a much-needed nitrogen snack for this plant.

INHABITANTS OF THE TROPICAL WORLDS

BALSA

FETCH THE NURSE!

When a hole appears in the canopy, the fast-growing **BALSA TREE**—also known as a "nurse tree"—helps to bandage the wound.

The balsa quickly sprouts its large leaves and shoots up, up, up, to shade the young seedlings below.

LEAFY LOOPHOLES

Many plants wage chemical warfare against insect attacks. At the first insect bite, some plants send up a chemical signal that tells the rest of the plant (or its neighbors) to beware of intruders.

Despite these defenses, **LEAF-CUTTER ANTS** have a cunning tactic, leaving plants none the wiser. With military precision, their chainsaw-like jaws vibrate one thousand times a second, cleanly slashing rather than munching a leaf, which makes it harder for plants to detect. The leaves are hard to stomach, but the ants have found a way to manage that: They use fungi to break down the fresh leaves into something they can eat.

Ants are able to remove up to 20 percent of the leaf before the plant realizes it's under attack.

JAMAICAN NETTLETREE LEAF

LEAF-CUTTER ANT

THE VAMPIRE

The **RAFFLESIA** is a vampire on Borneo's forest floor. It has no leaves for photosynthesis, so it burrows into climbers to suck food and water out of them. After months, the rafflesia's cabbage-shaped buds burst out of the climbers and reveal one of the world's biggest flowers. It's also known as the corpse flower, because these blooms reek of rotting flesh to attract flies (its favored pollinators).

RAFFLESIA

A STICKY SITUATION

Wounds in the forest canopy can let in invasive plants, such as the fast-growing **SILVERLEAF DESMODIUM**. This plant is covered in fine, hooked hairs along its stem that help it swamp the vegetation around it. The hairs also work like Velcro, ensnaring unsuspecting birds, mammals, and insects. In Madagascar, people have reported seeing up to forty frogs stuck in an area just three feet square (about the size of a bath towel).

THE STRUGGLE TO SEED

In the tropics, no sooner do seeds hit the ground than they're under siege from insects, mammals, and fungi. Fortunately, the **DIPTEROCARP TREES** in Borneo, the tallest trees in the tropics, have a plan. Every few years each tree puts all its energy into producing lots of seeds. They take flight, all at once, like mini helicopters.

There are simply too many for the **BEARDED PIG** to snatch, so some seeds will escape the snuffling snouts and grow into the next generation of giants.

DIPTEROCARP SEED

BEARDED PIG

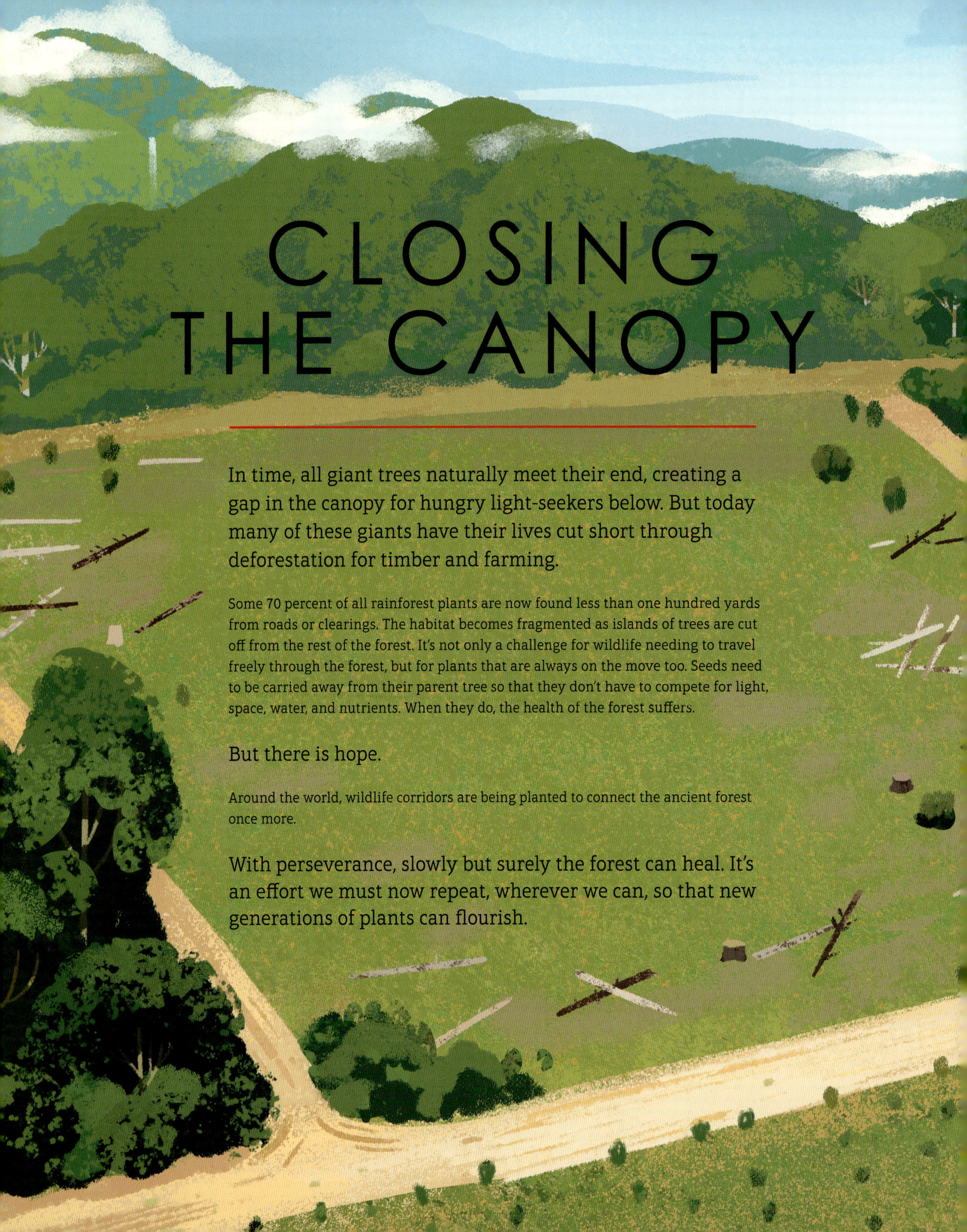

CLOSING THE CANOPY

In time, all giant trees naturally meet their end, creating a gap in the canopy for hungry light-seekers below. But today many of these giants have their lives cut short through deforestation for timber and farming.

Some 70 percent of all rainforest plants are now found less than one hundred yards from roads or clearings. The habitat becomes fragmented as islands of trees are cut off from the rest of the forest. It's not only a challenge for wildlife needing to travel freely through the forest, but for plants that are always on the move too. Seeds need to be carried away from their parent tree so that they don't have to compete for light, space, water, and nutrients. When they do, the health of the forest suffers.

But there is hope.

Around the world, wildlife corridors are being planted to connect the ancient forest once more.

With perseverance, slowly but surely the forest can heal. It's an effort we must now repeat, wherever we can, so that new generations of plants can flourish.

DESERT WORLDS

JUST ADD WATER

In the rainforest it's a battle for light, but in the desert, plants are fighting for something else: water. It makes survival here a very tricky business indeed. But on their never-ending quest to quench their thirst, some of the world's most persistent plants have clever tactics to find every last drop.

Some plants have no foliage, or grow smaller, thicker leaves that lose less water. Many plants stockpile water in barrel-like trunks. Others prefer to steal, intercepting water from their neighbors. And those plants that *have* water want to protect it from others, often in the most vicious possible way, like cacti with their needlelike spines.

Yet, for all the desert plants that steal, store, and secure their water, many others are content having a long snooze, waiting for the rains to come. *And it can be a very long wait.*

As the first raindrops fall over the Sonoran Desert in America, underground seeds that have been biding their time can finally flourish. The desert bursts into life with species such as the **MEXICAN GOLD POPPY** stretching for miles.

It's beautiful, but brief.

As the dry returns to the desert, the blooms fade, and dust storms called **HABOOBS** sweep up the sands.

The seeds are scattered to new corners of the desert, where they'll begin the long wait for rain once more.

STORIES
FROM THE DESERT WORLDS

SECURITY STAND BY. OVER.

In America's Great Basin Desert, the **WILD TOBACCO PLANT** is determined to protect its water-filled leaves. It creates a deadly poison—nicotine—which is toxic to most animals. It will, however, tolerate a visit from the **TOBACCO HAWK MOTH** to help with pollination.

The hawk moth stops by at night, lured by invisible chemicals released from the tobacco plant's flowers.

But the moth soon outstays its welcome, sneakily laying eggs that will hatch into leaf-chomping **CATERPILLARS**.

The caterpillars can tolerate nicotine, and they're hungry . . .

Luckily, the plant has just the snack—hairs covered in a delicious sugary treat (nicknamed "lethal lollipops").

The caterpillars greedily wriggle and nibble their way through their snack but . . .

. . . SNIFF, SNIFF . . . it gives them a body odor that smells delicious to nearby predators.

SNIFF, SNIFF.
And here they come . . .

The **BIG-EYED BUGS.**

The bugs fly to the rescue, dispatching the caterpillars and the unhatched moth eggs.

But wait, a few caterpillars have survived. And they're getting bigger. And bigger.

Soon the caterpillars' jaws are strong enough to chomp the leaves. At the first bite the tobacco plant sends a signal from leaf to leaf, alerting the entire plant.

Chemicals in the leaves give the caterpillar poop a special smell that attracts the tobacco plant's back-up crew . . .

The LIZARDS!

DON'T EVER CALL THE CHOLLA CACTUS CUTE

This plant might look cute and fuzzy, but what you're looking at are the **CHOLLA** *(choy-a)* **CACTUS'S** brutal barbs. Its arms produce buds, which are covered in barbs that plunge and hook into an animal as a painful reminder to stay away from the cholla's water-filled flesh.

But wait, what's this?

Something even cuter than a cholla has just scurried by. The **PACKRAT**: a fuzzy little furball with adorable eyes. But this cutie pie is, in fact, a master architect.

Though most animals give the cholla a wide berth, the packrat collects the cholla's prickly buds to build a fortress. It might get the odd prickle, but surprisingly the packrat is never skewered.

Inside its prickly fortress the packrat stashes stolen treasure—from silverware to false teeth—and raises its pups.

Better still, the cholla benefits from the packrat's help. Those buds not eaten can be knocked out of the nest, rolling away to find a new place to take root.

CACTUS COMPANIONS

The gigantic, treelike **SAGUARO CACTUS** of America's Sonoran Desert has a pleated, barrel-shaped trunk. This trunk can expand like an accordion to store over six tons of water (that's enough to fill a bath eighteen times). But it's only enough to last the saguaro a few months. And when the **GILA WOODPECKER** uses its long, pointed beak to hammer out the saguaro's trunk, the precious water begins to evaporate.

So the saguaro creates a thick sap that hardens into a boot-like shape, fixing the leak—and making a perfect home for the Gila.

Word spreads that it's a real HOOT living in the saguaro, so when the Gila moves on, the **ELF OWL** moves in. Over many years, the saguaro boot will house many different creatures, helping to make the Sonoran Desert one of the world's most diverse ecosystems.

INHABITANTS OF THE DESERT WORLDS

TUNNELING THROUGH TREES

In Zimbabwe, the mighty **BAOBAB** can live for thousands of years, able to survive droughts by holding thousands of liters of water in its trunk. But in some areas the baobabs are under pressure: There are too many hungry elephants in too small a space. The elephants tunnel through the baobab's soft bark to eat the spongy fiber inside—creating gaping holes that leave the oldest baobabs too weak to survive.

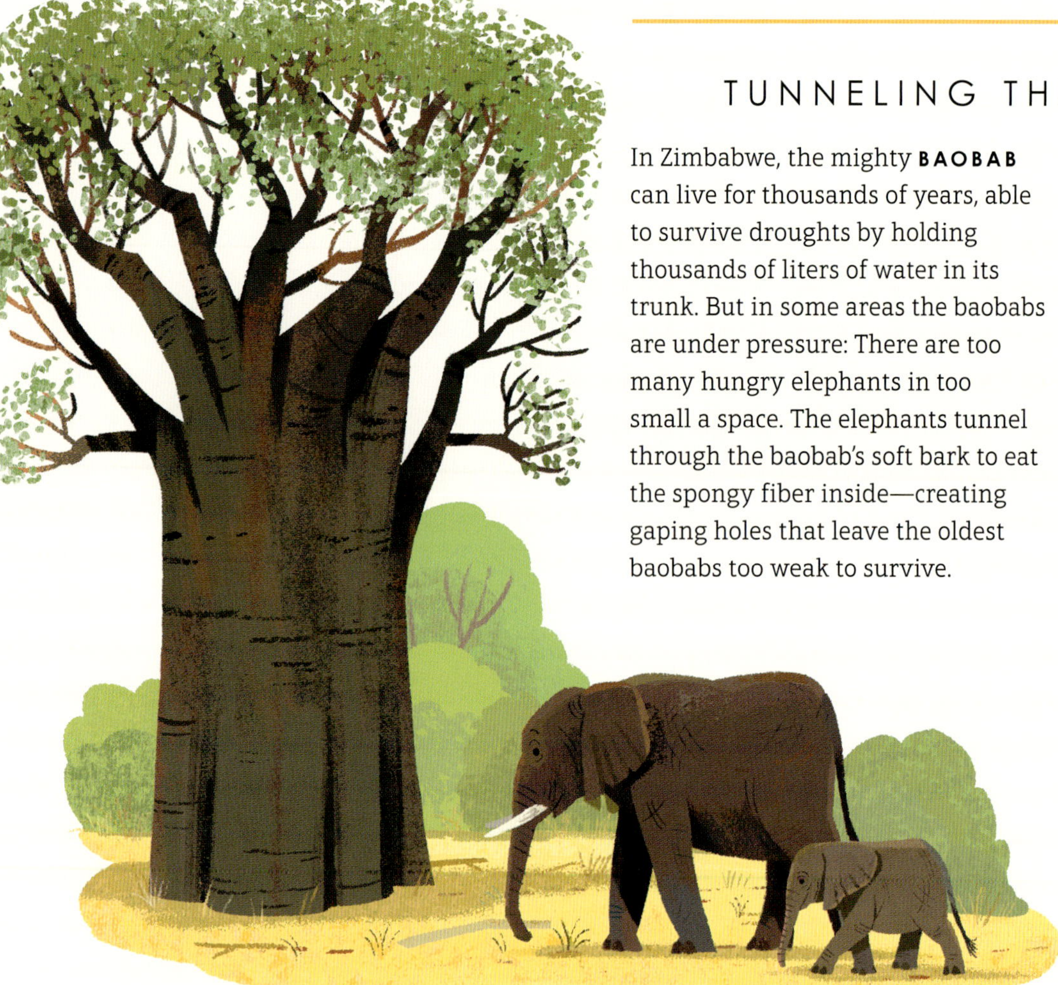

BAOBAB TREE

AFRICAN ELEPHANTS

JUST ADD POOP

Tiny Isla San Pedro Martir in the Gulf of California has been nicknamed Booby Island for its fabulously footed residents—many thousands of **BLUE-FOOTED** and **BROWN BOOBIES**.

CARDON CACTUS

But where there's a lot of boobies, there's also a lot of poop, making the soil virtually uninhabitable for plants . . . with the exception of the **CARDON CACTUS**. This is one of the densest cactus populations anywhere on Earth, and they thrive living among piles of poop and drawing water droplets from fog.

DESERT THIEVES

CACTUS MISTLETOE

The crafty **CACTUS MISTLETOE** resorts to stealing water from its neighbors. The **CHILEAN MOCKINGBIRD** eats the fruit of a mistletoe and poops the seed on to the spike of a **HEDGEHOG CACTUS**. There, the mistletoe grows and stretches out its **HAUSTORIUM**, a long fingerlike root, to burrow into the cactus. Inside the cactus the mistletoe grows bigger, and bigger, nourished by the cactus's water. Eventually it bursts out of the cactus with a beautiful red bloom.

CHILEAN MOCKINGBIRD

HEDGEHOG CACTUS

A SLOW SURVIVOR

In America's Mojave Desert, the **CREOSOTE BUSH** has learned to sip, not suck. In that way, it has grown ever so slowly, just thirty-two millimeters in half a century. That's barely three times longer than the average adult eyelash. Around 11,700 years old, this seemingly unremarkable circle of shrubs is one of the oldest plants on Earth, outliving mammoths and saber-toothed cats.

CACTUS MISTLETOE

SURVIVING IN THE SANDS

Desert plants are hardy survivors that have found extraordinary ways to adapt in Earth's harshest landscapes.

The Sonoran Desert's saguaro cactus is able to grow as tall as seven adult humans and weigh as much as seven bulls. It can live for up to two hundred years, withstanding just the right amount of heat, cold, and dry. It's a delicate balance that until now has worked in the saguaro's favor—and for other plants and animals too. The saguaro is a keystone species: It helps other wildlife to survive, and gives them food and shelter. But when they're young, saguaros rely on other species for their own survival.

Their lives begin as a seed, eaten by birds and then pooped out under "nurse trees" such as the mesquite or the paloverde, which shade the young cacti from the scorching heat of the day and give shelter in the bitter cold of the night.

But today the desert worlds are changing. Global temperatures are rising, and many nurse trees are being lost to human disturbance.

Among ten thousand fully grown saguaros, scientists have found just seventy young cacti. Their future, and that of the many species that depend on them, is at a tipping point. We can only hope that the seeds of these hardy desert survivors can make their way to more favorable conditions—not too hot, too cold, or too wet—so that the Sonoran Desert will be a sea of saguaros again.

WATER WORLDS

THEY CAME FROM THE DEEP

Some scientists believe that life on Earth may have begun around four billion years ago, huddled around a deep-sea vent. Here in our water worlds, marine life slowly flourished and spread. Among it were simple green algae that would play a BIG part in how plants conquered this bare Earth.

Studies suggest that around five hundred million years ago, algae might have gone on the march, leaving the sea to explore the world above water. Over time the algae evolved to survive in very damp places and then, *very slowly,* these new life-forms evolved to live in much drier places on land. They sprouted roots to reach down for water and nutrients, trunks and branches to reach up for life-giving sun, and flowers to reproduce. They created a land rich with food and oxygen, in an evolutionary leap forward for animal-kind.

But the plants weren't done yet.

In a dramatic U-turn, many plants went on the march once more, discarding their land adaptations and evolving new ones for an aquatic life. And life underwater is rarely straightforward. Water can lack nutrients and be a turbulent place to call home—plants have to either hold on or be swept away.

In the fast-flowing Caño Cristales River in Colombia, a water weed known as **MACARENIA** uses its roots (called holdfasts) to cling on. In the wet season, when water levels are high, the macarenia is busy growing at the bottom of the river. But when the water levels drop, sunlight and air help the macarenia to bloom in brilliant red. It casts river rainbows as its colorful leaves sway above the yellow sand and the black rocks in the river's blue waters.

In many corners of the planet, water plants spend their lives submerged. So when the waters recede, they peek above the surface to flower. Time is of the essence—they must attract a pollinator and set seed before the waters rise once more.

Welcome to the ever-changing water worlds.

STORIES FROM THE WATER WORLDS

ATTACK OF THE BLADDERWORT

High up on Venezuela's "Devil's Mountain," rain lashes the rocks, washing away soil and nutrients. Here, something sinister lurks . . . the murderous **BLADDERWORT**. Bladderworts are the largest group of carnivorous plants in the world. And this lean, green, eating machine has made the unsuspecting **BROMELIAD** its best friend.

Bromeliad leaves cup water to keep the plant hydrated, and these mini pools can house invertebrates, including **BEETLES**, **SPIDERS**, **TADPOLES**, and **SCORPIONS**.

But all is not safe at this watering hole. *A bladderwort is hunting.* Its long, tentacle-like stolons plunge into bromeliad pools to grow where the animals swim.

Along each stolon, pinhead-sized bladders grow—watertight, with a hairy trapdoor. An unsuspecting **INSECT LARVA** goes by.

Careful . . . it's a trap.

It comes closer. And closer.

The larva brushes past the bristled trapdoor and

WHOOSH!

It's sucked inside.

No predator attack in the plant kingdom is faster.

The larva peers out of its see-through prison cell as the bladderwort begins to digest it.

The trap is reset.

Another snack will swim by soon.

THE FUSSY FLYTRAP

The nutrient-poor, waterlogged soils in Green Swamp, North Carolina, can be unwelcoming for a flowering plant. *Yet something is stirring in the bog.*

Alien-like objects emerge from the ground—it's a world of tentacled slime traps (from **SUNDEWS**) and pit traps (from **PITCHER PLANTS**) as carnivorous plants capture their prey. Yet the queen of these hungry plants is undoubtedly the **VENUS FLYTRAP**.

Its hinged leaves resemble a toothy mouth, but it doesn't snap shut for just anyone. The crafty Venus flytrap only bothers with the biggest bites.

Something has touched one of the Venus flytrap's trigger hairs. It could be nothing. *Best to wait and see.*

Another hair is touched.
That's two strikes, so the mouth . . . shuts.

But even then, an escape route is left open. (The Venus flytrap doesn't waste its energy digesting tiny critters.)

Each further touch of the trigger hairs spells doom for its struggling prey. After another five touches the now fully sealed trap is flooded with acid so that digestion can begin.

5, 4, 3, 2, 1, TRAPPED!

For good.

WELCOME TO THE BEETLE BASH

In Brazil's Pantanal, the world's largest inland wetland, you'd think there was plenty of room for everyone. But the cloudy waters are no good for photosynthesis, so there's a dramatic battle for the best spots at the top. And in the fight for light, **GIANT WATER LILIES** use heavy force to command the water.

The lily launches a spiny bud from the depths. It's whirled like a weapon, appearing to swipe at the competition. Spiky leaves unfurl over six feet wide to smother anything in their path.

The giant water lily blankets the surface, its stem anchored in the muddy depths below. The huge, waxy leaves—more than twenty per plant—hog the sun, blotting out the light for everything growing below.

Having won the surface attack, the biggest bully in the pond can now reveal its bounteous bloom. Its flower bud shoots to the surface and, as the sun sets, it unfurls a white female flower. But this isn't any old flower. *This is the hottest spot in town.*

Inside the flower it's ten degrees warmer, and with a perfumed scent, the flower is irresistible to **SCARAB BEETLES**, already covered in pollen from visiting other lillies. The beetles turn out in droves, queuing forty at a time. As the party rages the flower shuts—it's a lock-in.

The beetles keep on partying well into the next day, mingling and eating. All the while, changes are afoot. Unknown to the beetles, the female flower has become male, turning pink in the process.

It's not until later that day when the lily is no longer producing its delicious aroma, that the flower opens. Now covered in pollen, the beetles fly into the night. The flower closes for the last time, its ovules fertilized, ready for the next season, as it slips under the surface.

As for the beetles . . . the party's not over. It's time to check out the newest lily hotspot just across the water.

INHABITANTS OF THE WATER WORLDS

ROLLING, ROLLING, ROLLING

Clumps of green **ALGAE** are scattered along the wintery shoreline of Japan's Lake Akan. As the ice melts, the algae is washed into the crystal-clear waters where its journey begins . . .

Here, it slowly grows, gently buffeted by the wind and waves, the constant motion spinning it into a perfect sphere known as a **MARIMO BALL**.

WHOOPER SWAN

Growing less than a centimeter a year, marimo balls can take one hundred years to grow to the size of footballs.

Some balls never make it beyond the hungry **WHOOPER SWANS**, but those that do roll deeper and deeper away from herbivores.

They huddle on the lakebed in mysterious colonies, thousands strong.

MARIMO BALLS

ANYONE FOR SALAD?

In South America's ponds and lakes, flotillas of **WATER LETTUCE** bob on the surface, their dangling roots sucking nutrients from the water. They look and feel like a fuzzy lettuce, but for the **HOATZIN** *(hoh-at-sin)*, these leaves make for one smelly salad. After a lettuce lunch, the hoatzin (also known as the stink bird) rests its paunch over a branch, and it gets to work fermenting the leaves.

It's a windy business—PARP!—resulting in a lot of stinky gas.

THE NEED TO SEED

Unlike animals, plants can't choose where to raise their young, so they have developed clever tactics to spread their seeds far and wide.

Some plants stack the reproductive odds in their favor by producing stacks of seeds, such as the **BULRUSH** (a wetland grasslike plant) that can hold 220,000 seeds.

Then, there are seeds that swim. In Brazil's Bonito Pools, fish are friends with **FIGS**! The migratory **PIRAPUTANGA FISH** have become handy seed couriers as they leap out of the water to pluck low-hanging fruit from the fig tree. In the days it takes the fish to digest their meal, they'll have swum upstream and will poop out the seeds into a new watery home.

THE DUCKWEED TAKEOVER

The world's smallest flowering plant, **WOLFFIA**, a member of the duckweed family, rarely produces flowers. Instead it reproduces by cloning itself. Left unchecked over four months, a single duckweed plant could multiply enough to cover Planet Earth!

SECRET GARDENS IN THE SEA

In our marine worlds, one group of flowering plants returned to the sea one hundred million years ago to live and flower under the waves—seagrasses. They can survive the salt, movement of the sea current, and even munching from hungry dugongs and green sea turtles, but the impact of our human world has taken its toll.

By 2000 almost half of our planet's seagrass meadows had disappeared because of disease, humans polluting the water, and building on top of coastal habitats. These meadows are home to many sea creatures and are a vital weapon against climate change. Seagrasses remove carbon from the atmosphere up to thirty-five times faster than tropical rainforests while also protecting our shorelines.

But the good news is that efforts to protect these meadows are helping to turn things around. For the first time in eighty years, seagrass meadows in Europe, America, and China are expanding.

Our secret gardens in the sea might flourish once more.

SEASONAL

SEASONS TO SCHEME

Plants that live between the poles and the tropics are ruled by four seasons. Some are evergreen and keep their leaves all year through. Others, like these deciduous trees, shed their leaves in the autumn to channel all their resources into their roots so they can survive the frozen winter ahead.

They stand bare, but not lifeless, and wait. As spring sunshine finally peeks through, it's like the sound of a starting pistol. Wakey, wakey, the forest is beginning to thaw. It's a race to feed, grow, and reproduce before the winter shutdown returns.

In North America, the **MAPLE TREES** stir as the sap begins to rise. Stored in their roots during the winter, this sugar and water solution now flows up to the plants' twigs so they can sprout leaves. But as the sap rises, something else is rising from its slumber . . . a sleepy **BLACK BEAR**, hungry from hibernation. It's time to make the most of the new season.

And it's not alone: **SAPSUCKERS** frantically hammer into the trunks to reach the sap. If the **HUMMINGBIRDS** are lucky, they can sneak a taste of the sap for themselves.

Time is short for many seasonal plants. In each season they have tasks to fulfill, using *secret tactics* in their race to reproduce. Some listen out for the pollinators coming so they can increase the concentration of their nectar, while others—such as thistles—use their long stems to wave the pollinators down.

Explore our seasonal habitats, and you'll find plants are collaborating, cheating, and chatting as they make the most of each season's opportunities.

THE WOOD WIDE WEB

Sssshhhh. Listen . . . All over the seasonal world, far beneath your feet . . . the plants and fungi are "*talking*."

Mushrooms, such as the **FLY AGARIC**, have a network of long, thin threads called **MYCELIUM** that reach under the soil for miles. These fungi threads plug into the end of tree roots, giving nutrients, while taking sugars for themselves.

But we're only now beginning to understand that this relationship between plants and fungi is much more complex. *It's driven by a desire to help each other.*

Just as the internet works by sending information through cables underground, the natural world has created its own communications network beneath the ground known as the Wood Wide Web. By eavesdropping on it, scientists have learned that older plants, known as "mother trees," use their deep roots to draw up nutrients to send out to shallow-rooted seedlings so that they can survive.

Mothers can also pick up distress signals from plants that are struggling and share nutrients in the same way. And dying trees can send their own nutrient stores to their healthy neighbors of the same, or even different, species.

Bonded by the mycelium of the fungi, the plants do more than swap food. Incredibly, they "talk," swapping information to support each other's survival. When one plant is under threat from leaf-eaters it sends out an SO

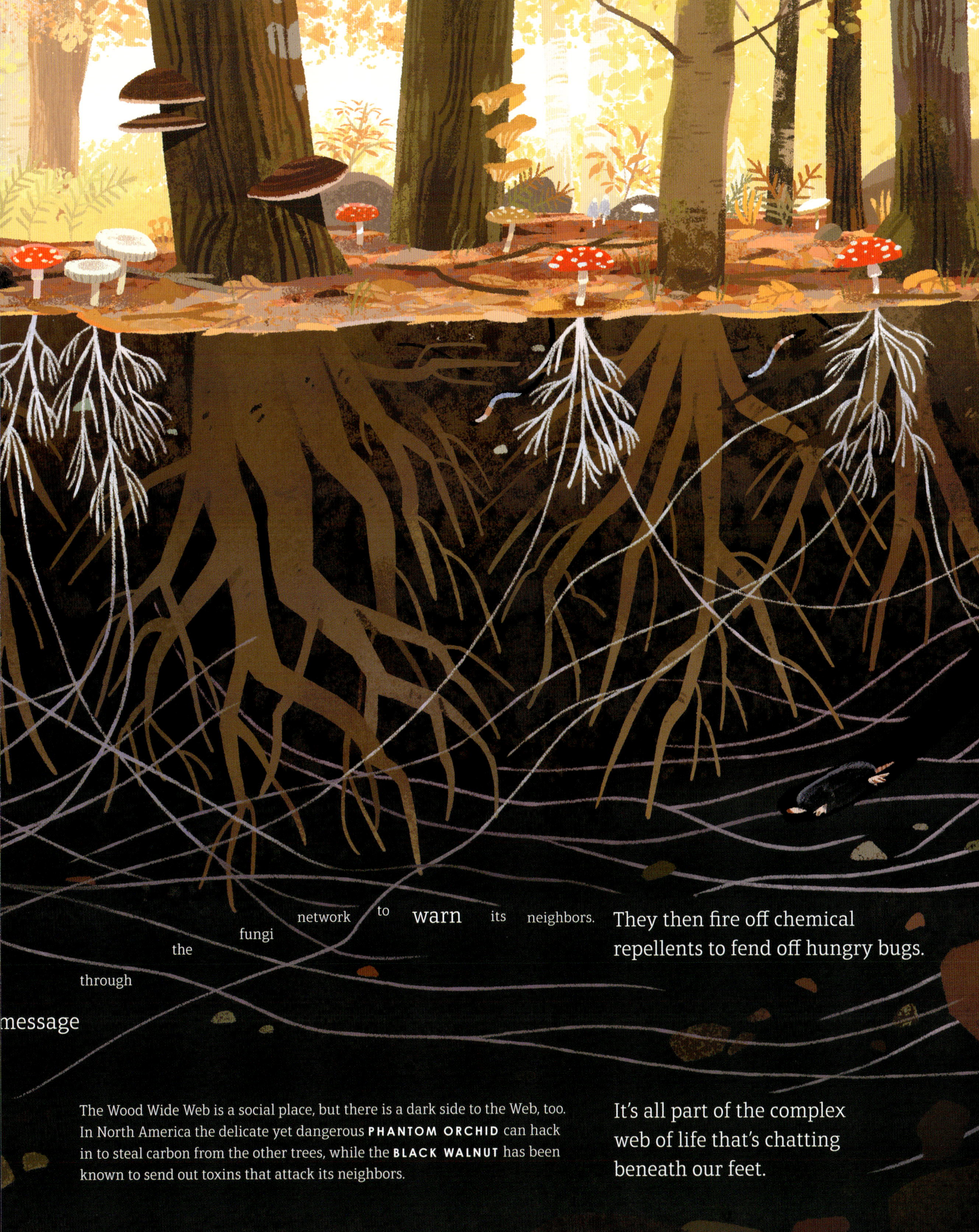

message through the fungi network to warn its neighbors. They then fire off chemical repellents to fend off hungry bugs.

The Wood Wide Web is a social place, but there is a dark side to the Web, too. In North America the delicate yet dangerous **PHANTOM ORCHID** can hack in to steal carbon from the other trees, while the **BLACK WALNUT** has been known to send out toxins that attack its neighbors.

It's all part of the complex web of life that's chatting beneath our feet.

INHABITANTS OF THE SEASONAL WORLDS

OAK

Not all plants are prepared when autumn's first frost falls. These ethereal **FROST FLOWERS** are sculpted from ice. They occur overnight when the plant stems release water that freezes in the air to form delicate ice petals. They're fragile and fleeting, gone by sun-up.

THEY LIKE TO MOVE IT, MOVE IT

Young **DAISIES** are eager spring sunseekers. In slow motion they're swaying back and forth as they follow the sun's journey east to west across the sky. This is **HELIOTROPISM**, and although it looks exhausting, it's the best way for young daisies to soak up maximum rays so they can grow.

There's a curfew: Strictly no swaying after nightfall when they close their petals to the elements.

At dawn, the daisies yawn open and turn their faces east for another day of sunseeking.

UNBE-LEAF-ABLE

Leaves come in all shapes and sizes.

DWARF WILLOW

The world's tiniest tree, the **DWARF WILLOW**, survives chilly Arctic conditions by staying close to the ground, its three-millimeter leaves as small as half a grain of rice. While the grand **AMERICAN SYCAMORE** has large maplelike leaves that can grow up to twenty centimeters wide—wider than an eight-year-old's hand.

High up in the canopy where sunshine is plentiful, leaves tend to be smaller (like those of an **OAK TREE**) compared to those further down in the shade (such as the **DOGWOOD**). Seasonal plants need their leaves to grow big enough to soak up sunshine for photosynthesis, but not so big they lose precious heat and are threatened by nighttime frosts.

DAZZLING DENS

From above, an **ARCTIC FOX'S** home looks like a summer oasis among the faded tundra—their deep, century-old dens carpeted in lush grass and wildflowers. Arctic foxes are accidental gardeners, as the nutrients from their poop, urine, and leftover dinner bones encourage grasses and flowers to grow. Their colorful dens can attract wildlife from across the tundra.

ARCTIC FOX

DECEITFUL DUNG

You'd think there's not a lot to be gained by smelling like a pile of poop, but South Africa's **CERATOCARYUM GRASS** *(sair-ah-toe-care-e-um)* knows otherwise.

Normally, an **ELAND ANTELOPE** has no sooner done its business, than busy little **DUNG BEETLES** scamper in to fight over the droppings, which they roll into their burrow to lay their eggs on.

So in summer, ceratocaryum grass produces round seeds that look—and smell—like an eland dropping.

The dung beetle will mistakenly roll it back to its burrow and, in doing so, plant the seed.

CERATOCARYUM
GRASS SEED

DUNG BEETLE

THERE BE GIANTS

The biggest living tree on our planet—the giant sequoia—reaches heights no dinosaur ever knew. Growing twenty-five stories high, these mighty evergreen trees need up to four thousand liters of water each day. With a lifespan of over three thousand years, they're one of the world's most successful seasonal trees. It's hard to imagine that anything could topple these giants. But today the giant sequoias are so terribly thirsty.

Giant sequoias were once found throughout the Northern Hemisphere but are now clustered in around seventy small groves on California's Sierra Nevada mountains. Spring snowmelt trickles down from the mountains, sustaining the sequoia through the summer until the autumn rains fall. But human-made climate change is making the summers longer—and hotter. All that sunshine means the giant sequoias are growing faster than ever before. But the bigger they get, the more they transpire and soon . . . the water runs out.

We've always believed that giant sequoias could survive anything, from drought and fire to disease. Now, desperate to keep every last drop of water, the sequoias are not only shedding their needles, but they're dropping entire branches, too.

But so long as we keep striving to curb climate change and stabilize the seasons, these beautiful giants will endure.

HUMANS

WE PLOWED PARADISE

Plants entwine our Green Planet and make it possible for all living things to call Earth home. But we don't always see them. What was the last animal you saw? Can you remember its color, or its name? Now, what about the last plant you saw, can you remember it too?

We don't always notice plants, something scientists call "plant blindness," which can mean we underappreciate the flora all around us.

Over many thousands of years our relationship with plants has changed. For a long time humans and plants evolved together, with many plant species thriving in the new environments we created. But in time we began to cultivate the plants that we liked best. We decided how some plants should behave, and bred them to become disease-resistant or more productive, or to have the showiest flowers.

In our quest to make it easier to produce our food, we've reduced biodiversity (the variety of life on Earth). In many places this has led to monoculture (single crop production) dominating the landscape, as we've singled out plant species to create a never-ending conveyor belt of food and timber.

Our single-mindedness has come at a price: When we reduce biodiversity, we weaken the whole ecosystem.

Today plants are disappearing fast—in two hundred fifty years, nearly six hundred plant species have gone extinct. The ripple effect for all of nature, including humans, is serious.

We rely on plants, and it's clear that they now rely on us too.

Fortunately, this Green Planet is home to people who are determined to look after it. And everyone can help.

STORIES FROM THE HUMAN WORLD

LIVING BRIDGES

One of the wettest places in the world, Meghalaya, is perched on a steep rock above India's Bengal Plains. Here, monsoon rains wash away anything not bolted down, creating torrents of mud—it's no place for humans. Yet through billowing waterfall mist, over raging rivers and across impassable ravines, extraordinary **FIG TREES** have extended a lifeline to the local Khasi people.

The fig trees send down a tangle of aerial roots to anchor in the slippery rock. Over the length of a human lifetime, the Khasi patiently weave these roots into living bridges. By respecting "tree time" the Khasi have created a kingdom in the canopy.

Bridges can stretch for more than fifty yards and exist for *many* centuries.

A testament to what can be done when we work with, not *against*, nature.

Ethiopia's misty highland plains are not an easy place to call home. The thin atmosphere makes it blisteringly hot by day and bone-chillingly cold by night. Only the strong survive. And the roughest, toughest resident on the plain? The humble **FESTUCA GRASS**.

Spiky silica stalks protect this tough grass against temperature extremes and munching mouths. This also makes it useful to humans: to stuff mattresses, weave ropes, and thatch roofs. Add to this a booming human population and the Festuca is fast disappearing. The grass is cleared to make way for crops and more livestock, leaving the ground dry and trampled.

But in a remote corner of the plateau is a grassy oasis.

Among the wavy Festuca grass, **"BLEEDING HEART" GELADA BABOONS** are dotted across the landscape, joined by ginger-furred **ETHIOPIAN WOLVES**.

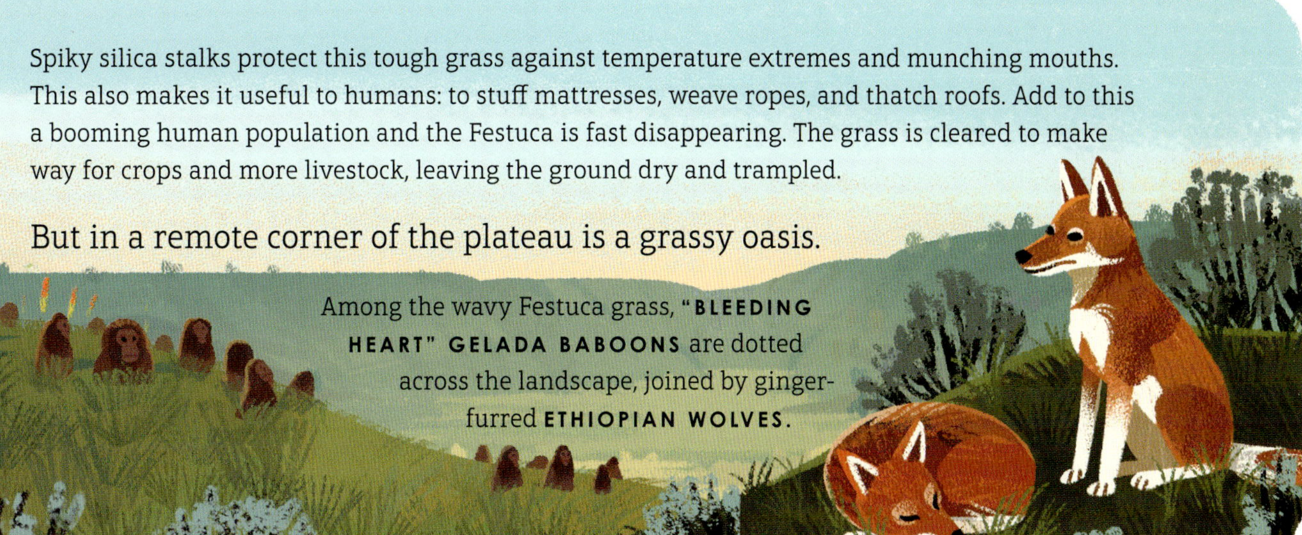

This is the Menz-Guassa Community Conservation Area, a biodiversity hotspot that's protected by the people.

Festuca is valuable to poachers, who creep through the grasslands with their scythes. But the Guassa people have protected it for over four hundred years, securing the health of this vital habitat.

INHABITANTS OF THE HUMAN WORLD

SELECTIVE SEEDERS

Plants often use animals to spread their seeds farther than they could manage on their own. For humans, collecting seeds scattered on the ground is a painstaking job, so through selective breeding, we've produced plants that seed in a more convenient way for us.

In the wild, when cereal crops such as **WHEAT**, **BARLEY**, and **RICE** ripen, they naturally shatter their seeds from the spike. We've now changed these crops to grow bigger seeds that stay on the spike until we're ready to harvest, then eat them.

In its quest for the perfect spot to seed, one wild plant does something extraordinary. Resembling a spiky mouse, the **WILD OAT** spikelet uses two bristle tails that spin against the ground so the plant can slowly "walk." It's looking for a rock to hide under or crack to drill into, so its seed can grow.

BEES FOR HIRE

The arid Central Valley of California grows around 82 percent of the world's **ALMONDS**. It's monoculture on a massive scale—a multibillion-dollar business and one that's very thirsty. The crop is so unnaturally large that there could never be enough native insects to pollinate it, so today's farmers must truck in **EUROPEAN HONEY BEES** to help get the job done.

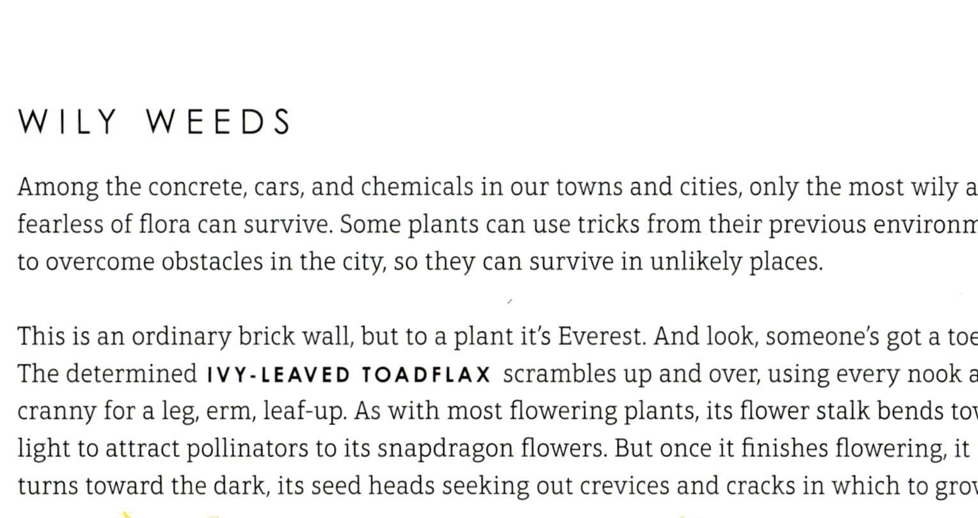

WILY WEEDS

Among the concrete, cars, and chemicals in our towns and cities, only the most wily and fearless of flora can survive. Some plants can use tricks from their previous environment to overcome obstacles in the city, so they can survive in unlikely places.

This is an ordinary brick wall, but to a plant it's Everest. And look, someone's got a toehold. The determined **IVY-LEAVED TOADFLAX** scrambles up and over, using every nook and cranny for a leg, erm, leaf-up. As with most flowering plants, its flower stalk bends toward the light to attract pollinators to its snapdragon flowers. But once it finishes flowering, it turns toward the dark, its seed heads seeking out crevices and cracks in which to grow.

While some weeds climb, others prefer to fly. The fluffy white seeds of the **SOW THISTLE** drift off on the wind, far from the parent plant.

Once we are down at the weeds' level, we begin to have a little sympathy for these fearless flora.

SOW THISTLE SEEDS

POLLINATORS UNDER PRESSURE

Around 40 percent of invertebrate pollinator species, especially bees and butterflies, now face extinction. This is due to the heavy use of insecticides on farms and in gardens, as well as a loss of habitat, which takes away their food sources. Thankfully, humans are beginning to realize how prized pollinators are for every mouthful of food we eat. We can change the way we farm and garden to help protect pollinators and the wild plants that support them.

BUMBLEBEE

HORNET

BLOWFLY

HUMMINGBIRD HAWKMOTH

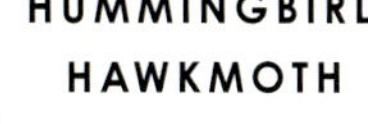

NATURE FINDS A WAY

Plants help to make this Green Planet a world that we can call home. Yet every day, more of our forests and grasslands are lost to human development—the world of plants shrinking as our human world grows.

Scientists estimate 77 percent of Earth's land outside Antarctica has been modified as we smother green with gray to cultivate our concrete jungles, and farm and mine the land. Plants that dare peek through, where we've decided they don't belong, are squirted with herbicides—whatever it takes to keep unwanted invaders out. But nowhere better illustrates the extraordinary resilience of plants than the concrete jungle. Even here nature sometimes finds a way to break through . . .

Along the crowded streets of Hong Kong, the banyan trees have rebelled. Grown from seeds dropped by birds or bats hundreds of years earlier, they've woven their way through the city's stone walls. Abandoned buildings are tangled in trees as nature stakes its claim on the city. It's evidence of our Green Planet's extraordinary resilience, that even in a concrete city, green can still grow.

And with a helping hand from humans, glorious gardens and thriving farms can bloom. We are the vital ingredient to ensure our Green Planet grows.

CALLING ALL PLANT PROTECTORS

For many centuries the world of people and the world of plants were compatible. But in recent decades the balance has shifted so that today around two in five plant species face extinction. There's still hope: Humans can learn how to correct their mistakes. From huge global projects to local action, plant protectors are helping our Green Planet to flourish once more.

LOCAL ACTION: SEEDBALLS

In Africa, charcoal is the fuel that powers peoples' lives. It's made from burning wood, which has led to millions of trees being cleared. But in the drylands of Kenya, communities have embraced "throw to grow" as a way to reforest their land. Seeds from the ancient **ACACIA TREE** are tucked inside a ball of waste charcoal. They're then lobbed from paragliders and helicopters, dropped by schoolchildren or grazing camels, and even fired from slingshots, as everyone does their bit to help restore the forests.

GLOBAL ACTION: SAFEGUARDING SEEDS

Each week, seeds are collected from all over the world and stored in seedbanks. The world's largest and most diverse seedbank is the Kew Millennium Seedbank in England. Here more than two billion seeds, from almost forty thousand species in 190 countries, are kept safe as a backup plan against future environmental catastrophes. As well as keeping seeds safe, Kew's scientists wander the world identifying new species and finding out what climate change means for them all.

INDIVIDUAL ACTION: PASSING IT ON

In 1994 Sebastião Salgado and Lélia Wanick Salgado inherited the family ranch in Brazil, but instead of finding the forested lands from Sebastiao's childhood, they discovered little more than a desert. The trees that were once part of the Atlantic rainforest had been cut down for timber, the soil eroded by livestock grazing, and the water had dried up.

Rather than giving up, the Salgados decided to carefully replant paradise.

1994

Tree planting began in 1999, and within two decades they had created a young forest, helping many species of animals to return to their natural habitat.

The Salgados set up the Instituto Terra, which has planted more than six million seedlings from over 290 native tree species. They're also passing on seedlings and everything they've learned to help a new generation of farmers so they, too, can green their corner of the planet once more.

2019

GO NATURAL

Herbicides and insecticides used to wipe out nuisance weeds and bugs are toxic to our Green Planet . . . and us! Without insects, what will the birds, amphibians, and reptiles eat? Who will pollinate the flowers? Chemicals from insecticides and herbicides can stay in plant roots and the soil for many years and end up in our food and water. You could see if your school can grow its own chemical-free fruit and vegetables—they're good for nature and great to eat.

GET PLANTING

Forests help to absorb carbon dioxide gas from factories, planes, and cars, and store it in their trunks and roots. Wetlands and grasslands also help to prevent our planet from getting any warmer by locking carbon into the soil. And the good news is that you can help too, by joining projects to replant previously cleared forests or to keep natural meadows and wetlands healthy.

Our lives, and the lives of plants, are perfectly entwined—we now depend on each other for survival.

IT'S NOT TOO LATE FOR US TO TURN OVER A NEW LEAF.

Humans and plants cannot only survive together . . . we could also thrive together on a shared Green Planet.